HETI Annual Membership Directory

2012

ISBN 978-0-9845490-5-4

THE FEDERATION OF HORSES IN EDUCATION AND THERAPY INTERNATIONAL A.I.S.B.L

SECRETARIAT
HETI
P.O. Box 134 • Damariscotta, ME 04543 • Tel: +44 207 1831 256
Email: office@HETIfederation.org • www.HETIfederation.org • www.facebook.com/HETI.international

ISBN 978-0-9845490-5-4

HETI Annual Membership Directory

2012

ISBN 978-0-9845490-5-4

Table of Contents

Letter from the President

Dear members of HETI,

Writing a foreword for the HETI Membership Directory is, for me, a humbling experience. Glancing through the pages of the last Directory, the scope of equine assisted activities and therapy around the world is really brought into full magnification. In the course of our daily lives we become cocooned in what is easy and familiar yet as I look at the programs around the world, I realize that there is so much more to do, to see, and to understand. Over the last couple of years, I have been many places and met many of you, but it is obvious, from this Directory, that there are many more organizations that I would like to work with and to learn from.

I wanted to share with you some of the ways that I have used the HETI Membership Directory – it is a very useful document!

- In my role as a college Professor teaching students to become Instructors, I use the Directory to broaden my students' vision of the breadth and depth of their field of study, and to easily contact other programs and organizations when searching for factual information.
- As a university Advisor, I am frequently looking for sites for internships for my students. The Directory puts the world at our fingertips for potential hosts.
- As a grant writer, I am able to use the Directory to help identify potential collaborators and partners – something that many funders are now requiring of their applicants.
- When speaking with parent groups or associations, I am frequently asked if equine assisted activities and therapy is well-known outside of a limited number of people. All I have to do is show them the Directory and they understand that this is something found the world over.

As I have said above, I use the Directory for a variety of reasons but perhaps one of the most important is to ensure that the use of equine assisted activities and therapy continues to expand in culturally acceptable, progressive and informed directions. When I meet someone new – I show them the Directory and encourage them to join HETI, as our strength comes from sharing knowledge and experiences.

I hope to meet new friends, renew old friendships and continue my journey in Taiwan for the XV International Congress – I hope you will join me!

Ann O'Shallie

President, HETI

Letter from the Executive Director

It is my pleasure to greet you again and send along the latest membership directory. In it you will find the rich tapestry that makes up our organization – 211 members in over 40 countries. Our membership continues to expand with enthusiastic additions and renewals. It is an uplifting experience to receive phone calls and emails daily about the work you are doing in equine therapy. Please use this directory as the resource it is intended to be, to spread the message about equine therapy and to network and communicate with one another.

I had a wonderful experience in Greece at the XIV International Congress! Time did not permit me to attend as many presentations as I would have liked but I met many old friends and made new ones while attending to HETI's exhibition booth. With a fruitful Congress accomplished, we can set our sights on the next Congress in Taiwan in 2015.

Some of HETI's goals for 2013 are:

- To continue to develop our journal, Scientific and Educational Journal of Therapeutic Riding into a first-rate peer-reviewed scientific journal
- To pursue education and training venues on a global scale. The message we receive from many countries is clear about the desire for this. We are continuing to investigate funding for these ventures, through grants and financial aid.
- The need for an internationally accepted terminology to facilitate research and education continues to be paramount. We believe this is achievable through continued dialogue and communication with all our members.

This directory is one of the fruits of the labors of our members. We send it to you hoping that it will inspire you in the year to come. Please stay in touch with me via email, the phone and of course our Facebook page.

With best wishes,

Gisela H. Rhodes
Executive Director/ Secretary, HETI

Our Sponsors

Petals, Inc., Australia
Uta Rindfleisch-Wu, Taiwan (in memory of Vernon Rhodes)
Betz Haartz, USA (in memory of Vernon Rhodes)
Norma Pearce, Australia
Pat Sayler, USA

Become a Sponsor!

HETI is at the heart of international Equine Assisted Therapy Education. Your sponsorship makes it possible!

- Support our members in spreading cutting edge research results through Newsletters, Publications and International Congresses.
- Help us promote the field of Equine Assisted Activities through literature, lectures and hands-on workshops.
- Support humane horsemanship methods to ensure that our partner, the equine, benefits and enjoys activities just as much as the humans.

Interested? Please contact Gisela Rhodes at office@HETIfederation.org. Sponsorship benefits include:

Platinum Sponsor: $25,000 and up

An all-expenses-covered invitation to an International Congress
Invitation to meet the President and the Executive Committee of HETI and attend all functions as the special guest of the President
Plus, all lower tier benefits

Bronze Sponsor: $200 - $999

Listing in HETI's Annual Membership Directory
Listing in frequent e-newsletter
Logo and link on HETI's website

Gold Sponsor: $10,000 - $24,999

Invitation to visit a HETI member program in the country of your choice
Attend a personal tour of the program
Plus, all lower tier benefits

Silver Sponsor: $1,000 - $9,999

Establish a special scholarship fund in your name
Display your logo at the HETI booth during the International Congress
Plus, all lower tier benefits

HETI (formerly FRDI) History

Past HETI Presidents:

Dorothee Ames, Canada;
Dr. Carl Klüwer, Germany;
Gillian Lawrence, Canada;
Octavia Brown, USA;
Gundula Hauser, Austria;
Pauline Perry, Australia

An International Congress is held every three years and offers an opportunity for all of us to get together, exchange ideas and celebrate our achievements in the world of equine assisted activities. Our past congresses are:

1974 I. International Congress in Paris, France
Reeducation par L'Equitation

1976 II. International Congress in Basel, Switzerland
Rehabilitation through Riding - Riding Therapy

1979 III. International Congress in Warwick, United Kingdom
Riding as a Sport for the Disabled

1982 IV. International Congress for Therapeutic Riding in Hamburg, Germany
Overview of the international situation in Therapeutic Riding

1985 V. International Congress in Milan, Italy
Rehabilitation through Riding

1988 VI. International Congress for Therapeutic Riding in Toronto, Canada
Therapeutic Riding and Scientific Research

1991 VII. International Congress for Th. R. on Aarhus, Denmark
An Update on Therapeutic Riding

1994 VIII. International Congress in Hamilton, New Zealand

HETI History (continued)

1997 IX. International Congress in Denver, USA
Riding the Winds of Progress

2000 X. International Congress in Angers, France
Cheval et Différences

2003 XI. International Congress in Budapest, Hungary
The Complex Influence of Therapeutic Riding

2006 XII. International Congress in Brasilia, Brazil
Contact between Friends

2009 XIII. International Congress in Münster, Germany
Horses for Body, Mind and Soul

2012 XIV. International Congress in Athens, Greece
Horses in Education and Therapy

Our next Congress will be:

2015 - XV. International Congress in Taiwan

June 22-25, 2015, at
Aletheia University,
Danshui, New Taipei City,
TAIWAN

Apply to the Mary Selway-Swift Development Fund!

Mary Selway-Swift

This fund was established from a bequest to HETI (FRDI) from the estate of the late Mary Selway-Swift.

Mary was involved in the field of therapeutic riding over many years - in the United Kingdom and in South-East Asia (specifically in Hong Kong and Singapore). It was Mary's wish that this fund be used to promote and improve the educational opportunities, training and quality of riding therapy through the development of therapeutic riding programs internationally. She was particularly concerned for those in developing countries or in places which have limited access to training opportunities. Any application needs to be supported by a HETI Full Member and payments from the fund will be made on the recommendation of the HETI Executive Committee.

For an application please contact the office at:

HETI
P.O. Box 134
Damariscotta, ME 04553, USA
Email: office@HETIfederation.org

Members are encouraged to promote the Mary Selway-Swift Development Fund among friends and interested people/businesses. Donations to the fund should be sent to the Office.

Donate to HETI!

Looking for a practical way to make a donation to the Federation of Horses in Education and Therapy International? Please consider the following:

Order flowers to send to your friends and loved ones!

When you order your flowers through Petals Network, a portion of your purchase goes toward HETI's fundraising efforts. Petals Network have a beautiful range to choose from.... bouquets, arrangements, roses.... and they do same-day delivery throughout Australia and the UK and next-day delivery to more than 70 other countries throughout the world. Please visit the HETI website for a direct link or use **www.petals.com.au/flowers/AI9093auauauden.html**

Learn how to make a rope halter!

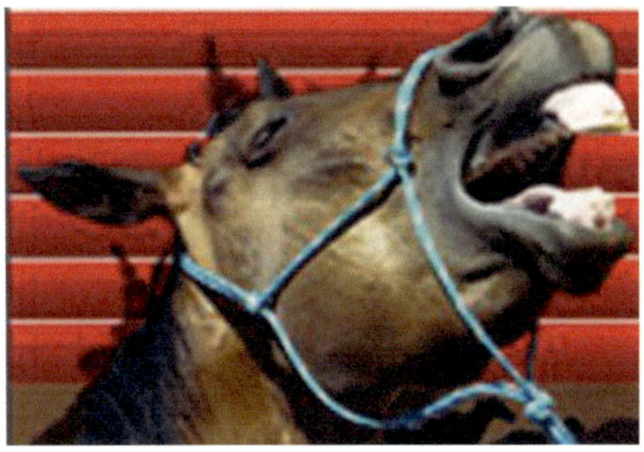

Tying your own rope halter is fast becoming a lost art. This video will show you how to make the halter with the Fiador knot which is the correct knot for making a halter, but often omitted as it is thought to be very difficult to tie. Following the simple directions on the video will make it easy for you to learn. How does this support HETI? Easy! Visit the HETI website for a direct link or type in **www.rope-halter.com**. Then make a donation of your choice ($5 minimum) and download the video or, with a $10 minimum, you will receive a DVD. Rope halters are an awesome personalized gift to any horse lover!

Thank you to our Advertisers

This directory is partially funded by our advertisers. We would like to extend our heartfelt gratitude to the following for supporting this publication:

- Clover Lodge
- FN Verlag
- Francesco Zipedia Astorga
- HHRF
- Lapo
- PATH International
- Special Equestrians

Their advertisements appear on the pages that follow.

The HETI Directory is distributed to all members, advertisers, and sponsors in over forty-five countries; it's our calling card. Its reach extends to almost anyone who's influential in the field of equine-assisted activities. If you would like your organization included with an advertisement, consider the upcoming edition.

For rates and placement in the 2013 Membership Directory please contact Gisela Rhodes at the HETI office:

P.O. Box 134
Damariscotta, ME 04543
Email: office@HETIfederation.org

ASTORGA A.C.

PSYCHOPEDAGOGICAL HIPPOTHERAPY AND WELFARE DEVELOPMENT CENTRE ASTORGA announces the following products:

1.

A pedagogical board for hippotherapy. Attached to a girth, it is adjustable to any size horse and includes simple stirrups.

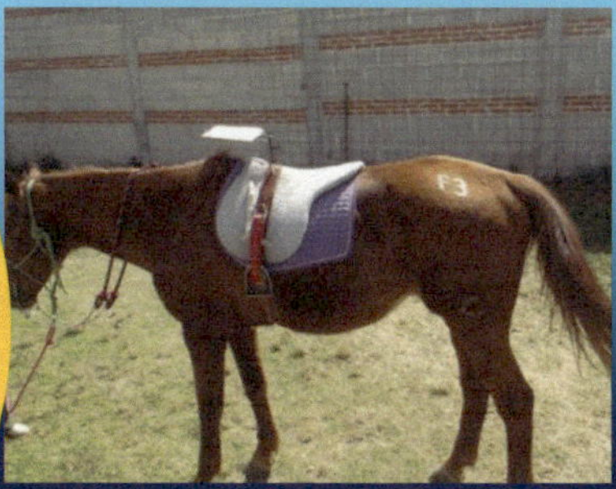

2.

A kit with specific training materials for therapeutic riding, including suggested activities.

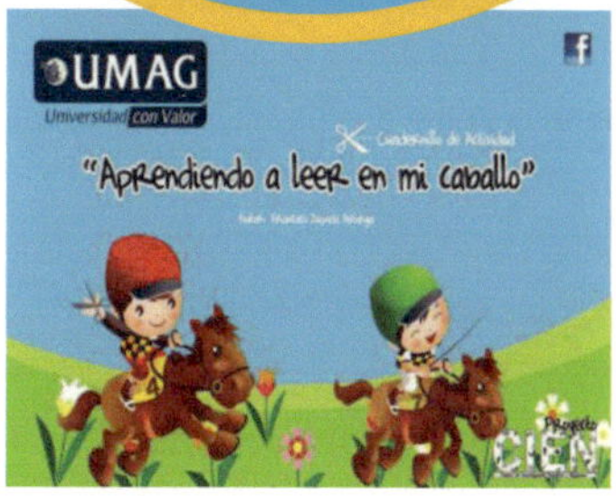

***Plus**... a student workbook to help with the acquisition of Spanish reading and writing skills.*

For inquiries please contact:
L.E.E. Francisco Zepeda Astorga • Calle Estado de Tlaxcala #98
Colonia la Santísima San Martin Texmelucan Puebla México 74021
zepedaastorga@hotmail.com • Office (01248) 48 4 10 02 • Personal 22 25 13 35 83

Have you ever tried to keep six students engaged and happy... and only had one horse?

Hello Bob! + 49 Other Interactive Vaulting Games is written for vaulting instructors and coaches who are looking for creative activities besides the vaulting compulsories. It is also an equally valuable tool for instructors of able-bodied and therapeutic riding sessions. In plain, often humorous language Gisela H. Rhodes shares her secrets accumulated during fourteen years of experience with therapeutic riding. Each game features specific information on skills addressed, set-up and materials, and detailed instructions on how to play the game safely for horse and student.

A gem of a book!

*"**Finally**, we have a collection of activities that Gisela has used for many years in her vaulting workshops. Some are very simple and others much more challenging - all proven to be very effective, educational and motivating. All of us at Windrush are delighted that this much anticipated book is finally available as a teaching tool. Both, students and instructors will be inspired and will benefit from this work."*

Amanda Hogan, Executive Director of Windrush Farm Therapeutic Equitation, PATH INTL. Master Instructor

Hello Bob! + 49 Other Interactive Vaulting Games

ISBN: 978-0-9801325-2-6

$22.95

Shipping:
$5 plus $1 per book

Gisela H. Rhodes, M.Ed., is a PATH Intl. (NARHA) Interactive Vaulting and Advanced Therapeutic Riding Instructor, as well as a Centered Riding© Instructor Level 1. She has served as the chair of the PATH Vaulting committee, as an EFMHA Board member and is currently executive director of HETI (FRDI). Her workshops, lectures and clinics are sought after, both nationally and internationally. Born in Germany, she lives at Fair Tide Farm in Newcastle, Maine.

Special Equestrians, Inc. • 220 River Road • Newcastle, Maine 04553 • www.fairtidefarm.com • (207) 563-3974 • info@fairtidefarm.com

HIPPOTHERAPY SADDLE "ELISA 2"

"LAPO" ASSOCIATION, IN COLLABORATION WITH "PRESTIGE ITALIA" SADDLERY, HAS REVISED ITS PREVIOUS HYPPOTHERAPY SADDLE, ELISA 1.

THE NEW "HIPPOTHERAPY SADDLE ELISA 2" IS CAPABLE OF MEETING THE SPECIAL REQUIREMENTS OF SAFETY AND ABOVE ALL, A MORE CORRECT POSTURE. SEAT, FLAP, CANTLE OF "ELISA 2" ARE CONFORMED IN SUCH A WAY AS TO PROMOTE A BETTER CORRECT SEAT AND TO FACILITATE PELVIC POSITION, TRUNK-STRAIGHTENING AND BALANCE. THE SPECIAL TRAPEZOID-SHAPED EXTRACTABLE HANDLE, WITH A PARTICULAR INCLINATION, THAT ALLOWS A SECURE GRIP AND A CORRECT POSITION OF TRUNK AND SHOULDERS, IS NOW PROVIDED BY A NEW SAFETY DEVICE.

To place an order please refer to the manufacturer, Prestige Italia:
www.prestigeitaly.com
info@prestigeitaly.com

Technical Features:

The "Elisa 2" saddle has a medium deep seat, with round cantle. It comes with flocked panels; it is offered with a flaps' forward projection standard/ + 2cm or + 4cm, depending on the different pathologies. Elisa 2 is provided with a set of medium front and back blocks with Velcro attachments to help riders find the correct position that best suits their needs.

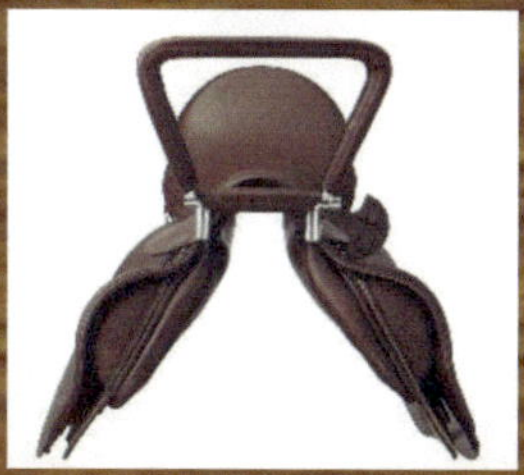

Associazione "Lapo" ONLUS Sezione "L.R.E." - Via A. Susini, 14 - 50125 Firenze (Italy)
Phone: +39 347 5398994 - Fax +39 055 4298432
associazionelapo@gmail.com - www.associazione-lapo.it

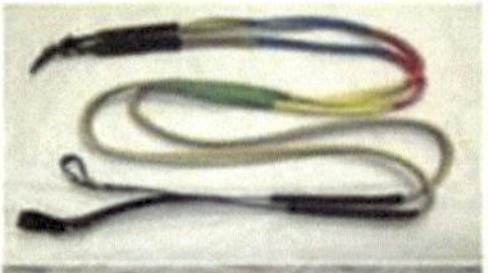

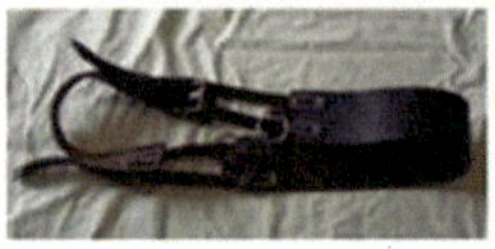

Gary and Sally Watson
620 Roycrofts Rd. Boorcan, Victoria, 3265
Australia
Email: cloverlodge@bigpond.com
www. cloverlodgeequestriansupplies.com
Ph. 61 419 593 914 or Ph. 61 419 593 960

2012 HETI Annual Membership Directory

Full Members

ACPTR - Association of Chartered Physiotherapists in Therapeutic Riding

Golland Farm
Golland Lane
Burrington
Devon EX37-9JP
UNITED KINGDOM
Contact: Helen Mathie, ACPTR International Liaison,
Phone: +44 (0)777 958 7472
Email: helenmathie@btinternet.com

ACPTR is a professional clinical interest group of the Chartered Society of Physiotherapists in the United Kingdom. Membership of ACPTR is not restricted only to Physiotherapists.

The function of ACPTR is to promote hippotherapy (horse assisted physiotherapy treatment) and best practice by physiotherapists working within a therapeutic riding environment.

ACPTR provides a post graduate course for therapists who choose to use the horse and its movement therapeutically and for Physiotherapists to qualify in Hippotherapy. It also organises workshops, study days and a biennial conference.

AEDEQ - Asociación Española de Equinoterapias

Calle Calderón 71, 2º, 5ª
Sabadell 08201
SPAIN
Contact: Lorenzo Lucarelli
Phone: 00 34 686331750
Email: lucarellil@hotmail.com
Website: www.aedeq.org

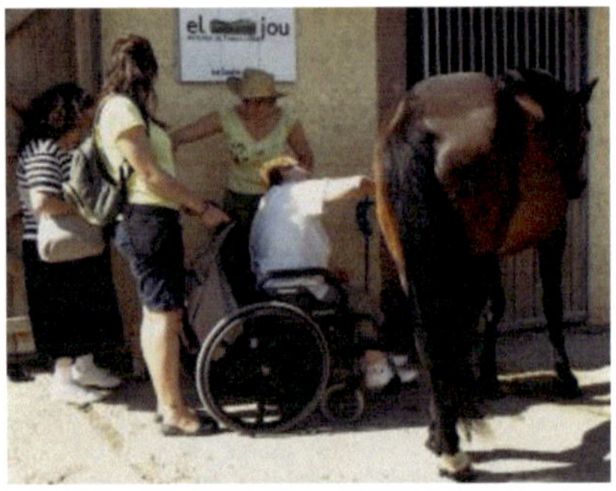

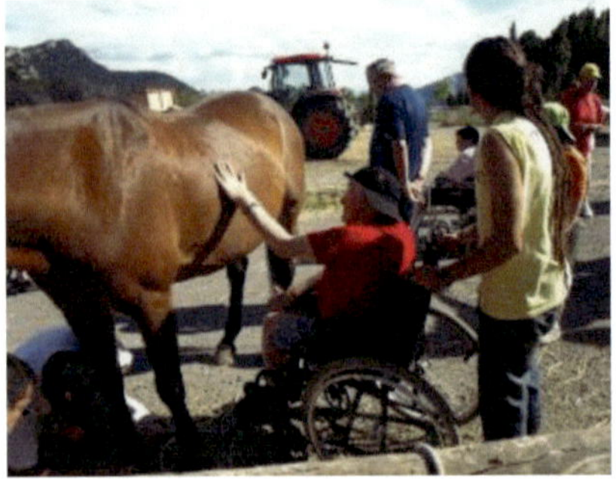

AEDEQ, the Spanish association for Therapeutic Riding has provided hippotherapy and equine assisted psychotherapy for several years and recently moved to a new stable. Since the organization's inception, it has been active in therapeutic riding as well as in leisure, sports and educational activities assisted by horses. The association is now in the process of organizing new services and incorporating them into their offerings:

- research the legal, economic, ethic, logistic, and social aspects of equine therapies
- organize various courses for therapists and instructors, at all levels, from initial to master
- incorporate therapeutic driving and vaulting
- prepare riders for adapted equestrian sports

AEDEQ is also active in promoting the REDEQUI cooperation project which is an international network of pedagogic centers providing therapeutic riding. The REDEQUI cooperation project uses tourism as a source of income to support the therapeutic team and to extend accessibility to therapeutic riding to those who cannot afford it otherwise.

AHA - American Hippotherapy Association

PO Box 2014
Fort Collins, CO 80524
USA
Contact: Jacqueline Tiley, Executive Director
Phone: (570) 224 4505
Email: ahaexecutivedirector@gmail.com
Website: www.americanhippotherapyassociation.org

American Hippotherapy Association Inc. (AHA, Inc.) promotes the use of the movement of the horse as a treatment strategy in physical, occupational and speech-language therapy sessions for people living with disabilities. Hippotherapy has been shown to improve muscle tone, balance, posture, coordination, motor development as well as emotional well-being.

AHA consists of medical professionals (physical, occupational and speech-language therapists) and others who are interested in the use of equine movement as a treatment strategy. Its philosophy includes the following:

- The organized movement of the horse is an effective treatment strategy used to improve function in patients with neuromusculoskeletal dysfunction.
- A licensed physical therapist, occupational therapist or speech-language pathologist with specialized training in the principles of hippotherapy and horsemanship is responsible for directing the treatment team during a session using hippotherapy as a treatment strategy.
- A hippotherapy session requires a carefully trained team consisting of: patient, therapist, horse expert, horse and sidewalkers.
- Equine Assisted Therapy, used by a medical professional, is part of a comprehensive treatment plan that integrates gains into functional activities.
- Horses used in Equine Assisted Therapy should have quality of movement, appropriate temperament and ongoing specialized training. All horses should be treated humanely at all times.
- Clinicians providing Equine Assisted Therapy are to maintain a high level of professional competence by participation in related continuing education activities. They must follow the standards of practice and code of ethics for their discipline.
- More research should be done to continue to validate the efficacy and effectiveness of hippotherapy and other Equine Assisted Therapies.

AMET - Mexican Association of Therapeutic Riding

www.equinoterapia.net/amet
amet@equinoterapia.net

Balcones de la Colina 25
Fracc. Balcones
Queretaro 76140
MEXICO
Contact: Blanca Mayol
Phone: 442 214 3906
Email: amet.mexico@gmail.com
equinoterapia.amet@hotmail.com
Website: www.equinoterapia.net

AMET has been serving children with disabilities for over 36 years and represents Mexico's equine therapy centers by integrating, training, certifying, and promoting them internationally.

AMET's mission is to provide hippotherapy for children, adolescents and adults with physical and mental disabilities, behavioral issues and / or learning disabilities as well as victims of accidents. The association seeks to contribute to a better understanding of the needs and intrinsic value of persons with disabilities and to promote better relations among people with disabilities and the rest of society.
Furthermore, AMET trains and assists other institutions interested in creating new Equine Centers and it coordinates, certifies and regulates the establishment and operation of those centers. AMET also represent and supports them, both to increase their international knowledge and to receive financial support.

AMET has given rise to more than 150 new centers in Mexico and assisted 14 municipalities in their state. Presently, they offer 94 Equinotherapy courses, covering topics ranging from therapy for small children, music therapy, finger painting to substance abuse, familial violence, Rett syndrome, autism and neurodevelopmental and other conditions that can be treated with equine therapy. Additionally, they have 11 courses in neurohabilitation and hippotherapy and just added to their curriculum Natural Horsemanship.

ANDE-BRASIL - Associação Nacional de Equoterapia

Granja do Torto
70620-200 Brasília DF
BRAZIL
Contact: Lelio de Castro Cirillo
Phone: (55) (61) 34687092
Fax: (55) (61)3468-8486
Email: ande@equoterapia.org.br
Website: www.equoterapia.org.br

Since its foundation in 1989, The National Association of Equotherapy, ANDE-BRASIL, has focused its work on the dissemination of technical and scientific knowledge within the Therapeutic/Health, Educational and Equitation areas for professionals from Brazil and other countries.

The teaching program of ANDE-BRASIL centers on basic and advanced level courses in Equotherapy for professionals with a university degree and it specializes in university courses on the post-graduate level. A basic course in Equitation for graduated professionals with a high school is also offered.

ANDE-BRASIL has also organized the following conferences:

- 1991- Brasília, 144 participants- I Course and Meeting on Equotherapy
- 1999- Brasilia, 300 participants- I Brazilian Congress of Equotherapy
- 2002- São Paulo, 320 participants- II Brazilian Congress
- 2004- Bahia, 368 participants- I Iberian-American Congress of Equotherapy
- 2006- Brasilia, 450 participants from 38 Countries- XII International FRDI Congress
- 2008- Paraná, 500 participants- I Latin American Congress of Equotherapy
- 2011- V Brazilian Congress
- 2011- II Iberian- American Congress on Equotherapy.

ANIRE - Italian National Association of Therapeutic Riding

Via Bartolomeo D'Alviano 7
Milano 20146
ITALY
Contact: Daniele Nicolas Citterio
Phone: 39 03 3927 63674
Email: info@cnranire.it
Website: www.cnranire.it

ANIRE is the national association which introduced riding rehabilitation to Italy in 1976 and it is still the only recognized association of its kind. It deals with rehabilitation or riding therapy, and regards riding from a purely rehabilitative angle as in helping people achieve greater self-sufficiency. However, this does not rule out taking part in competitive sports activities. ANIRE was founded with the aim of rehabilitating people with physical and mental challenges with the help of horses and of promoting riding for recreational and sports purposes.

ANIRE also promotes scientific research, organizes and promotes riding rehabilitation in Italy, and trains specialists in riding therapy.

Since its foundation, the association's membership has grown to 10,000, encompassing 5,000 users/customers, 1,980 specialist sector workers (650 of which are currently active) and 170 centres spread over 20 regions, serving 12,000 patients with over 350 horses. Increasingly, centres run by municipalities, local health centres, regional governments and traditional rehabilitation centres, such as hospitals, enter agreements with ANIRE.

ANIRE was conferred the Milan Provincial Government Gold Medal and received a Certificate of Merit from its local municipality.

ATE - Association Therapie Equestre

13, Rue Mausereck
L-3934 Mondercange
LUXEMBOURG
Contact: Christiane Betz
Phone: 00352-55 26 15
Fax: 00352-26 55 00 65
Email: secretariat@atelux.lu
Website: www.atelux.lu

A.T.E.
ASSOCIATION
THERAPIE
EQUESTRE

Created in March 1982, ATE has been renting a small equestrian centre since 2004 where eight horses, three little ponies and two donkeys are housed. The association includes three physiotherapists with a certificate in hippotherapy and fifteen aids who lead the horses and help secure the patients, sometimes working with 3-4 horses simultaneously. Presently, the centre serves 106 clients.

Clients are children and adults with physical and mental challenges. In the educational and social part of their program, instructors work with clients with attention deficit disorder, sensory integration disorders, or behaviour issues.

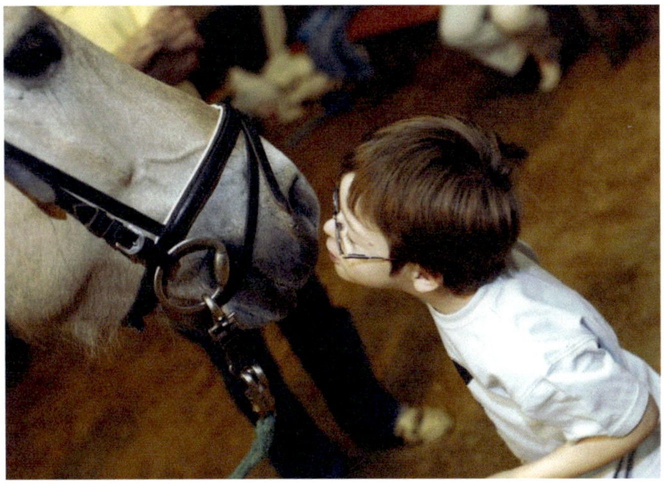

ATE works with clients individually or in groups of two, three or five, offering psychomotor exercises in form of vaulting or plays with psychomotor material in time and space. Clients learn to ride in groups with four horses until they are able to be integrated in a riding club outside.

CanTRA - Canadian Therapeutic Riding Association

5420 Hwy. 6 North, Suite 11, R.R.#5
Guelph, Ontario N1H 6J2
CANADA
Contact: Donna Naylor
Phone: 519-767-0700 Fax: 519-767-0435
Email: ctra@golden.net Website: www.cantra.ca

CanTRA's mission is "to promote the benefits of therapeutic riding throughout Canada, by raising public awareness, working closely with the medical profession, providing continuing education, maintaining high standards for Therapeutic Riding Instructor Certification and accrediting therapeutic riding centres."

CanTRA is the only national governing body for therapeutic riding and is recognized by the FRDI and Equine Canada. Nearly 100 member centres operate across Canada, working with children and adults who encompass a wide range of special needs.

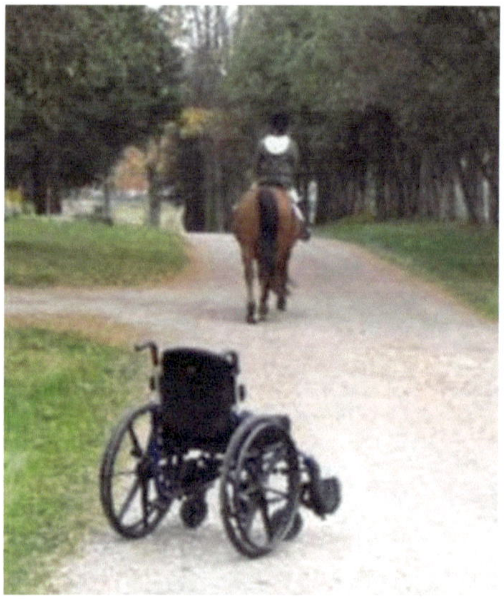

CanTRA has been a national association since 1980 and delivers programs which include

- an Internationally recognized **Instructor Certification** system with a high quality of service delivery
- an **Accreditation** system for therapeutic riding centres to ensure safe and high operating standards
- **Hippotherapy** training workshops for therapists
- opportunities for **Education** through newsletters, clinics, workshops and national conferences
- a national **Insurance Policy** available to member centres.

CanTRA also works to build knowledge and awareness of the mental and physical benefits using the horse as facilitator by establishing partnerships between certified equestrian personnel, therapists, and mental health professionals.

CNRAFA ONLUS - Italian National Centre of Research, Therapy and Training

Via Bartolomeo D'Alviano 7
Milano 20146
ITALY
Contact: Daniele Nicolas Citterio
Phone: 39 03 3927 63674
Email: info@cnranire.it
Website: www.cnranire.it

Acknowledged by decree of the President of the Italian Republic as the national Italian association of horse therapy and of sporting and recreational riding for people with disabilities, CNRAFA is certified according to the international parameters. A provider of the Ministry of Health for continuous medical education, the organization has an agreement with the Ministry of Education, Research and University and holds a convention with Mondino, a mental health clinic for the evolutional age belonging to the University of Pavia.

CNRAFA includes:

POLYVALENT CENTRE:
Horse therapy; Psychiatric area; Equestrian rehabilitation; Cognitive area ; Re-instatement; Working re-instatement; Sporting re-instatement. 148 patients, 298 treatments a week, 1.138 treatments a month, 11.380 treatments a year.

POLYVALENT TEAM:
Consultant in: Neuropsychiatrics; Physiatrics; Sports Medicine; Psychologist ; Physiotherapist; Motory sciences graduate; Motory sciences and techniques of the motory preventive activities graduate; Teacher; Instructor of equestrian sports; Nurse; Vet.

THE STRUCTURE:
Manège; Office; Waiting room; Dressing room; Saddlery; Stable; Human and veterinary infirmary; Record office; Didactics room; Farrier workshops; Stable and grooming workshops; Maintenance workshops; Film library workshops; Computing workshops; Gardening workshop.

CTEA - Chinese Taipei Equestrian Association

20 Chu-Lun St., Room 808
Taipei
REPUBLIC OF CHINA (TAIWAN)
Contact: Uta Rindfleisch-Wu
Phone: 8862-8771-1508
Fax: 8862-2778-3740
Email: ctea@ctea.org.tw
Website: www.ctea.org.tw

CTEA was founded in 1973. The main tasks of the association are to enhance the quality of horsemanship in Taiwan, to train coaches of different levels, to organize equestrian competitions and to choose riders for international competitions. CTEA has around 200 personal members and 30 group members. Taiwan is an island where horses were not indigenous and horse sports therefore do not have any traditions. English style riding is the common riding style in Taiwan and jumping competitions draw the most participants.
In 2002, when Para Equestrian Sports became a member of the FEI-family, this part of Therapeutic Riding was introduced to CTEA and in March 2008 the first PE Competition was held as part of a CTEA-Dressage Competition. Since 2009 CTEA got a director for PE-Sports.

Riders with disabilities joining able bodied dressage competitions have helped a lot to change the perception of the abilities of people with physical and mental challenges in Taiwan.

Several of the group members, mostly riding clubs, provide Therapeutic Riding for their clientele or are even specialized in this field, like the Therapeutic Riding Centre of Taiwan, which provides related information in Chinese through CTEA's magazine "Horses Whisper" and organizes the annual "Hope Cup", a RDA riding competition.

DKThR - Deutsches Kuratorium für Therapeutisches Reiten E.V.

Freiherr-von-Langen-Str. 8a
48231 Warendorf
GERMANY
Contact: Almut Schlingenkötter
Phone: 0049 (0) 2581/927919-0
Fax: 0049 (0) 2581/927919-9
Email: DKThR@fn-dokr.de
Website: www.dkthr.de

Deutsches Kuratorium für Therapeutisches Reiten e.V.

With nearly 3000 members and extensive connections with various professional associations, the German Association for Therapeutic Riding (DKThR) plays a leading role in therapeutic riding in Germany. As the professional association for all occupational groups and institutions, DKThR is responsible for the promotion and support of all professionals, hospitals, therapy centres, equestrian clubs and schools associated with therapeutic riding.

An important task is the development of a scientific basis and promotion of scientific publications as is the cooperation with government and private purchasers. Its international role is reflected in cooperation with FRDI and FATP. DKThR freely shares its know-how with other countries and furthers the exchange of information.

During the past 40 years DKThR has also established itself in the field of education and training of professionals and, in cooperation with the NF, developed quality management and quality standards. DKThR offers professional education in four fields of Therapeutic Riding:

- Hippotherapy
- Psycho-educated vaulting and riding
- Riding as sport for people with disabilities
- Occupational Therapy with the horse

Finally, DKThR supports the high-performance sport for disabled in Para-Equestrian –Dressage-Driving and is a member of the Special Olympics.

ELETHIP - Hellenic Association for Therapeutic Riding

18 Karaiskaki Str.
TK 19400 Koropi, Athens
GREECE
Contact: Irene Kapari
Phone: 0030 210 66 28 361 & 0030 6977 20 11 86
Fax : 0030 210 66 28 361
Email : info@elethip.gr
Website : www.elethip.gr

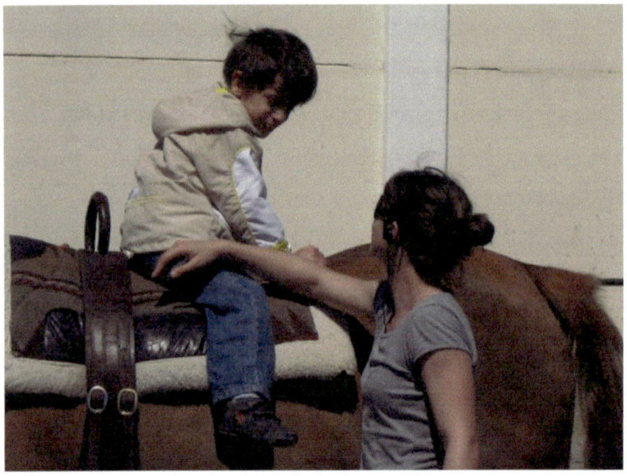

ELETHIP, the Hellenic Association for Therapeutic Riding, is a non-profit organization in Greece that for the past eight years has been providing all types of equine assisted activities to children and adults with motor, sensory, mental or behavioral problems.

The therapeutic team includes five specially trained horses with fine movement and character, as well as a team of experts in various fields, such as physiotherapists, occupational and speech therapists, psychologists and riding instructors.

The association participates in scientific conferences with papers and presentations and is involved in research in various fields. ELETHIP also organizes international seminars to train more therapists in therapeutic riding.

In the area of sports, ELETHIP sends many athletes to the Special Olympics Games and is looking forward to participating in Para-equestrian competitions in the near future.

Equestrian Federation of Finland

Radiokatu 20
00093 SLU – Helsinki
FINLAND
Contact: Merja Alasjärvi
Phone: 3580 - 9 – 15 82318
Fax: 358 - 9 - 1496864
Email: merja.alasjarvi@ratsastus.fi
Website: www.ratsastus.fi

The Equestrian Federation of Finland has included Riding for Disabled in the Federation since 1992. It provides training and education through specific courses for assistants, instructors and teachers. It also organizes competitions which are integrated into the dressage competition.

FCA - Fundacion Caballo Amigo

Lope de Vega No 36 5D
28014 Madrid
SPAIN
Contact: Jose Angels
Phone: 34 91 361 1463
FAX: 34 91 724 1125
Email: gestion@caballoamigo.org
Website: www.caballoamigo.com

In 1998, a group of professionals in medicine, education and sports founded the F.C.A. with the purpose of developing equestrian activities for disabled people in Spain. The foundation's activities focus on four main areas:

Therapy
Includes therapy, education, enjoyment, and social integration for people with disabilities with horses and riding. FCA is presently working with disabled riders with a variety of different disabilities.

Education
Courses in therapeutic riding are designed to train professionals from medicine, education and sports. FCA also offers a two year program in conjunction with the Complutense University of Madrid and the University of Bovigny (France).

Expansion
Includes various activities of the foundation with major emphasis on the scientific investigation of therapeutic riding.

Sport
FCA is responsible for the development of para-equestrian activities at the Spanish Equestrian Federation.

FCRT - Fortune Centre of Riding Therapy

Avon Tyrrell
Bransgore
Christchurch, New Forest, BH23 8EE
UNITED KINGDOM
Contact: Mrs. Jennifer Dixon-Clegg
Phone: 44 1425 673297
Fax: 44 1425 674320
Email: enquiries@fcrt.ac.uk
Website: www.fortunecentre.org

The Fortune Centre of Riding Therapy (Registered Charity Number: 1045352) is a place that provides education and therapy through horses for people that are more motivated by being with horses than anything else. Weekly sessions are run for local children with a variety of disabilities. A day care provision exists for local people with special needs who wish to focus on working with horses. FCRT provides work experience placements for allied professionals usually during their professional training in physiotherapy, occupational therapy, teaching or nursing.

The FCRT's main activity is as an Independent Specialist Provider for young people, aged 16-25, with learning difficulties and disabilities whose educational needs cannot be met in government provision. The residential course *Further Education Through Horsemastership* (FETH) runs over 3 years for 45 students at any one time from anywhere in the UK and occasionally from overseas.

Fédération Nationale Handi cheval

Glen Gouttes
26 Route du Pre Madame
Royeres 87400
FRANCE
Contact: Patrick Mc Laughlin
Phone: 33 555 69 21 98
Email: mclaughlin.patrick@akeonet.com
Website: www.handicheval.asso.fr

FENTAC - Fédération Nationale de Thérapie avec le Cheval

12 Allee du Parc
89170 St. Fargeau
FRANCE
Contact: Dominique Darques
Phone: 33 86-74-0378
Email: darques.dominique@orange.fr
Website: www.fentac.org

The National Federation of Therapy with the Horse is a non-profit organization created in 1986. FENTAC aims to:

- bring together groups and individual practitioners of therapeutic riding
- develop and oversee enforcement of specific guidelines for Therapy with the Horse
- promote research regarding the theoretical and practical aspects of this therapy
- support other organizations associated with Therapy with the Horse
- take part in meetings and conferences in France and abroad which relate to therapeutic riding

FISE - Italian Equestrian Federation

Viale Tiziano 74
Rome 00196
ITALY
Contact: Dr. STEFANIA CERINO
Tel: 39 06 36858369
Fax: 39 06 36858797
Email: riabilitazione@fise.it
Website: www.fise.it

FISE

The Italian Equestrian Sports Federation (FISE) was founded in Rome in 1926 and, approved by the Italian Olympic Committee and the Ministry of Culture, the Federation is the only authorized organization to regulate equestrian activities in Italy whether they are of an educational, athletic, recreational or training nature.

FISE has a specific Department for Therapeutic Riding. Its goal is to develop opportunities for people with disabilities so they can attain treatment through Therapeutic Riding.

Every year, FISE organizes special training courses for riding instructors and therapists in various areas of social health. The Therapeutic Riding Department also conducts special research in the mental health field, especially on the use of therapeutic riding in psychiatric rehabilitation.

Finally, the Therapeutic Riding Department also organizes congresses, seminars, and workshops to improve and promote Therapeutic Riding.

FN - Fédération Equestre Nationale / Deutsche Reiterliche Vereinigung e.V

Freiherr-von-Langen-Straße 13
48231 Warendorf
GERMANY
Contact: Soenke Lauterbach
Phone: 49 (0) 2581/6362-0
Fax: 49 (0) 2581/62144
Email: fn@fn-dokr.de
Website: www.pferd-aktuell.de

Established in 1905, the German Equestrian Federation (FN) is the leading association for all horse breeders, riders and those involved in the driving and vaulting sport in Germany. With some 7.700 member-associations and 750.000 personal members, the FN is the world's largest international equestrian federation and an important promoter of therapeutic riding in Germany.
It supports the following areas:

- professional and amateur sport
- horse breeding
- horse keeping
- animal welfare
- nature protection and nature conservation
- the treasuring of the cultural asset "horse"

The FN is responsible for all questions related to the horse, including breeding, horse maintenance, horse equipment, horse training and the use of the horse by human beings. In order to guarantee a high quality, FN regularly issues uniform guidelines and regulations, such as the APO. The APO also contains the regulations for the training program for specialized personnel in the area of therapeutic riding, which is used by the German Curatorship for Therapeutic Riding (DKThR) as the basis of its educational program. The German Curatorship is a member of the German Equestrian Federation and partners with the Federation in the field of education, labeling and qualification of riding and therapy facilities for the disabled.

The German Equestrian Federation supported the implementation of the XIII International Congress for Therapeutic Riding in 2009, in Munster.

FPG - Federatie Paardrijden Gehandicapten

Wekeromseweg 12
6816 VS Arnhem
THE NETHERLANDS
Contact: Dieuwke Papma
Phone: 0031-6-46745415 (Tuesdays and Thursdays)
Email: info@verenigingfpg.nl
Website: www.verenigingfpg.nl

Since 1968, FPG is the Dutch organization of riding for people with disabilities. It is a certified fundraising organization. FPG has seventy-five members, all local groups which offer horse riding lessons on a non-profit basis, but working on a professional level. For a list of members, please look at www.verenigingfpg.nl/ledenmaneges.php.

The Objectives of FPG are to provide quality and safe riding opportunities for persons with disabilities, for recreational, sportive as well as therapeutic reasons. Activities include:

- education of riding instructors (qualification level 3)
- education in vaulting for disabled
- refresher courses for instructors
- courses for volunteers
- programs on safety- and risk management in FPG riding schools
- development of quality and safety standards for EAT
- integration of activities with the National Equine Sports Organization, KNHS, and the Federation of Riding Schools, FNRS.

For 2010 FPG hopes to

- support the foundation of a national organization of professional equine assisted therapists.
- stimulate more scientific investigations into the effects of EAT
- support development of an EAT-course for therapists
- get all members onto the list of institutes certified as safe riding schools.

FTR - Faggruppen for Terapiridning

Norsk Fysioterapeut Forbund
Post Boks 2704 St. Hanshaugen
0131 Oslo
NORWAY
Contact: Eileen Strand
Phone: 47 66 8077
Email: eileenaa@online.no
Website: www.fysio.no/ORGANISASJON/Faggrupper/Terapiridning

Faggruppen for Terapiridning in Norway (FTR) is one of 12 professionally organised groups within the Norwegian Physiotherapy Society (NFF). Comprised of all those physiotherapists interested in or qualified to treat patients with therapeutic riding, it has approximately 85 members.

Therapeutic riding started in Norway in 1953 at Elsbeth Bødtkers "Ponystallen" near Oslo. She also founded the forerunner to FTR in 1973, established funding from the Health and Social Services budget and organised the first International Congress at her Stall in 1973.

Therapeutic Riding (TR) today is partially funded by the Dept. of Health; patients are refunded a total of 30 sessions per year. In addition they must pay a fee to cover the full cost of treatment which varies from centre to centre. Most work with groups of three to eight patients. In order to qualify to work with therapeutic riding, physiotherapists must pass an examination in horsemanship, a course and exam organised by the Norwegian Horse Centre. In addition, they must be able to document their own riding skills to level B dressage. Finally two courses are arranged by NFF in Therapeutic Riding – level 1 and 2.

FTR's goals for the future are to encourage research, promote network building within their field, and secure co-operation and exchange of teaching methods in Scandinavia and abroad.

Georgian Medical Federation of Therapeutic Riding

29 Ave Chavchavadze
Tbilisi 179
GEORGIA
Contact: Dimitri Tsverava
Phone: 99 532 231593/996868
Email: dodosokhadze@yahoo.com

In Georgia, scientific research in the various activities associated with therapeutic riding begun in the sixties and was directly connected to the name of Professor Dimitri Tsverava. For the first time, the methods of therapeutic riding were developed and introduced during the treatment of:

- cardiac ischemic diseases
- information neurosis
- hyperactive syndrome of attention deficit
- cerebral palsy of children
- scoliosis
- osteochondrosis
- coxarthrosis
- imperfect osteoporosis

The Academy of the Georgian Medical Federation of Therapeutic Riding offers post-diploma education courses and a total of eight doctoral theses have been defended at the Academy. These courses as well as symposia and conferences on the issues of therapeutic riding are systematically conducted to help physicians and methodologist-instructor specialize and to raise their level of qualification. During the two International conferences held in 2006 and 2007 the Federation welcomed the President of FRDI, Professor Gundula Hauser. She gave a very interesting report and chaired one of the sessions. Presently, the Academy's specialists are preparing a fundamental book on therapeutic riding to be published in cooperation with FRDI.

GUL - God Unlimited

Gateways House
London Road
Shrewton, Wiltshire SP3 4DL
UNITED KINGDOM
Contact: Dr Celia C Grummitt
Phone: 01980 621712
Email: office@god-unlimited.org
Website: www.god-unlimited.org

God Unlimited, Gul, is an Outdoor Therapy Centre in England situated on the edge of Salisbury Plain - a huge natural resource of unspoilt countryside which we use for walking, riding, cycling and navigation training. Gul is a registered Alternative Provider of Education in Wiltshire, allowing us to take school-refusing young people as well as those currently excluded from the education system.

Several of Gul's staff are dual qualified in Equine and Medical fields. Our CEO, Dr Celia Grummitt, is a local GP commissioner and the Mental Health Lead for Wiltshire as well as one of our trek leaders. Our Senior Therapist, Mrs Ruth Freeman, is a physiotherapist with a Master's Degree in Rehabilitation and is the house mother of our residential site.

We regularly provide Equine Assisted Therapy sessions for both adults and young people with a variety of physical, learning, emotional and mental disabilities.

People of all faiths and no faith are equally welcome to Gul.

HANDHIPPO - Belgische Nationale Vereniging van Therapeuten Gespecialiseerd in Hippotherapie / Union Nationale Belge des Thérapeutes Spécialisés en Hippothérapie

Rue de Beriot, 33/Maatschappelijke zetel: de Beriotstraat, 33
1210 Bruxelles-Brussel
BELGIUM
Contact: Marie-Thérèse Kuypers
Phone: 32/ 498 48 32 44
Email : handhippo@skynet.be
Website: http://users.skynet.be/handhippo

Founded in 1974, the national Belgian non-profit organization Handhippo is composed of therapists specialising in therapeutic riding. Their definition of therapy as a medical treatment in the broader sense covers all aspects of the patient, both physical and mental, and the horse represents everything that it brings by his presence, his relationship, his contact which is both rewarding and special.

Therapy is given under prescription or on medical indications.

Handhippo's goal is to improve the credibility of everyone's work and to present a coherent image of the activities associated with therapeutic riding to the different professions, the media and the public.

Hest og Helse

Vækerøveien 207
0750 Oslo
NORWAY
Contact: Mr. Kjell Myhre, Secretary General
Chair: Mrs. Anne Kristin Sydnes
Phone: +47 93 44 12 40
Email : kjell.myhre@nhest.no
Website: www.hestoghelse.no

Hest og Helse, Horse and Health, is a voluntary Norwegian national organization which was established in 1971. Currently they have about 300 members – individuals and businesses – and a number of well-qualified course instructors with extensive experience.

Hest og Helse's extensive training portfolio deals with the horse in health care and the exciting intersection of horses and human health. It includes:

- horses in psychiatry and social medicine
- disability management
- training in horse therapy
- management of children and adolescents

The organization's purpose is to promote the use of horses in the health context. Hest og Helse strives to actively contribute to the optimal quality in the use of horses to improve the physical, mental and social health of its clients.

MAGYAR LOVASTERÁPIA SZÖVETSÉG

Hungarian Riding for the Disabled Federation

Terez Krt 24
Budapest 1066
HUNGARY
Contact: Peter Edvi
Tel : 361 331 0906
Email : edvi@GYERMEKMENTO.HU
Website: www.lovasterapia.hu

The Hungarian Riding for the Disabled Federation was founded in 1997 with the patronage and support of the International Children's' Safety Service. At the time, no nationally acknowledged and accredited organization existed to provide courses in therapeutic riding.

The first training was held by Sister Chiara Hatton Hall and seven English specialists. The following years Prof. Dr. Med. Carl Klüwer, Octavia Brown, Anita Shkedi, Dr. Danielle Citterio, Peter Holzmüller, Susanne von Dietz, Isabelle V. Neumann-Cosel, Kurt Lehmann, Willem van Lieshout, Reinhard Tötchinger, Gundula Hauser and Sharon Herz-Freilich held training sessions.

The Hungarian Ministry of Education acknowledged and accredited the Hungarian Riding for the Disabled Federation's training in 2000. The Federation provides training for four different specializations:

- Teachers of children with special needs
- Physiotherapists
- Psychologists
- Para-sport trainers

The Federation is also involved in helping the Hungarian Parliament and ministries prepare laws pertaining to therapeutic riding, such as riding services in Hungary and the Hungarian sport law decree which affects para-sport. Since 2005, the Federation organizes each year a Competition for Disabled Riders, together with the International (CDI***) and National Dressage Competition (CDN). These events have become very popular and help promote the integration of people with special needs. Since therapeutic riding and para-sport riding are not supported by the Hungarian social security, most of the members work part time and people involved in the administration of the Federation carry out their duties on a voluntary basis. They wish to thank the excellent international trainers who have supported them in their task.

KATH – Korean Academy of Therapeutic Riding (RDA - Riding for the Disabled Association Korea)

1 Bugok-dong
Gunpo-si, Gyenggi-do, 435-737
SOUTH KOREA
Contact: Amy Ji Young Lee
Phone: 82-31-461-6637
Email: ji02.lee@samsung.com
Website: http://horse.samsung.com/eng/rd/rdsamsung.html

KATH (RDA- Korea) was founded in 2003 and registered as a full member of HETI in 2004. Its founding mission was to develop programs for people with disabilities that provide opportunities to ride and to help them attain the benefits to their health and well being that can be gained through therapeutic riding.

The purposes of KATH are:
1) To promote therapeutic riding throughout the country.
2) To facilitate collaboration between different organizations with therapeutic riding programs and whose objectives are charitable, scientific and educational.
3) To encourage medical and paramedical therapists, sports experts, and educators to specialize in applying therapeutic riding techniques within their field.
4) To develop standards of practice for therapeutic riding programs.

The organization is based at the Samsung RD center and cooperates with other centers to achieve the goals outlined above.

LAPO - Italian Association of Families, Bodies and Professions against Neurological and Psychiatric Diseases in Childhood and Adolescence

Via Antonio Susini n.14
50125 Firenze
ITALY
Contact: Prof. Anna Pasquinelli
Phone: 39-55-4298431; 416774
Fax: 39-55.4298432
Email: pasquinelli@neuro.unifi.it
Website: www.associazione-lapo.it

The Italian Association of Families, Bodies and Professions against Neurological and Psychiatric Diseases in Childhood and Adolescence (LAPO) was co-founded by Prof. Massimo Papini and Prof. Anna Pasquinelli. Its activities within the area of Therapeutic Riding focus specifically on the "Section for Study, Updating, Research in TR and Sport Riding for the Disabled". LAPO is responsible for organisation, information, professional advice, and supervision of therapeutic riding centres in Italy and maintains contacts with professional associations, public institutions, administrative and legislative bodies, and the National Equestrian Sport Federation. The association organizes a congress, conferences and workshops on therapeutic riding.

In collaboration with the Chair of Child Neuropsychiatry of the University of Florence LAPO developed the *University Curriculum in TR for Post-Graduate Physicians in Child Neuropsychiatry* (1995) and, in 1996, founded the Italian *Study Group on TR,* a multi-professional group with scientific goals which meets twice a year. LAPO also developed a *Theoretical/Practical Training Course in TR* in 1997 and a *University Master in TR* in 2002/2003.

LAPO promotes and contributes to the study, research and scientific experimentation on therapeutic riding and has, in collaboration with the University, developed a *TR Neuromotor Disorders Assessment Scale* for the quantitative assessment of the results on horseback and on the ground as well as a definition of the guidelines for therapeutic riding activities in a therapeutic riding centre. Published books include *Practical Principles of Therapeutic Riding* (1996) and a *Handbook of Therapeutic Riding: Principles, Methodology, Organization* (2010). LAPO also provide a bibliography on therapeutic riding and technical aids for riding with disabled riders.

NFTR - National Federation of Therapeutic Riding

Weschnikowskoje House 1, 3, App56
Moscow 111402
RUSSIA
Contact: Alexander Ivanovich Denisenkov
Phone: 8-903-187-49-42
Email: rider960@yandex.ru
Website:

NFTR is an alliance of people and organizations devoted to the development and implementation of therapeutic riding and the appreciation of its prominent position among other rehabilitation activities in Russia. The main tasks of the Federation include information sharing and implementation of methods; training of specialists in therapeutic riding; coordination of member activities, as well as representation and protection of their interests; building and maintenance of training and informational databases for therapeutic riding; and cooperation with Russian and International organizations.

NFTR consists of specialists from more than 130 Russian Cities, where therapeutic riding was selected as a means to assist individuals with health problems.

Together with the team of “Our Sunny World” NFTR, as a founding organization, assists in the training of specialists in Russia and CIS. It develops and publishes teaching manuals for use in therapeutic riding and for the organization of rehabilitation centers. From 1997 to 2002 NFTR, together with Georgian Medical Academy, graduated 122 specialists certified as “Rehabilitation Horse Riding Instructor for Physical Rehabilitation”. From 2003 to 2009, in collaboration with the Department of Preventive Medicine Faculty PK MR RUPF, about 280 specialists were graduated with the degree “Specialists in Rehabilitation Horse Riding”.

NFTR is an acting member of the Federation of Horses in Education and Therapy International (HETI) and an active participant in European and international conferences on therapeutic riding.

NZRDA - New Zealand Riding for the Disabled Association

Po Box 58110
Whitby
Porirua, 5245
New Zealand
Contact: Guy Ockenden
Phone: 64 4 234 6090
Fax: 64 4 234 6094
Email: admin@rda.org.nz
Website: www.rda.org.nz

New Zealand Riding for the Disabled Association Inc.

Riding for the Disabled was first established in New Zealand 50 years ago. NZRDA is the umbrella organisation for its 56 member groups located throughout New Zealand. They provide over 3,000 children and adults with an opportunity for enjoyment, therapy, rehabilitation, freedom, friendship, laughter, and the achievement of independence and confidence with the help of the horse.

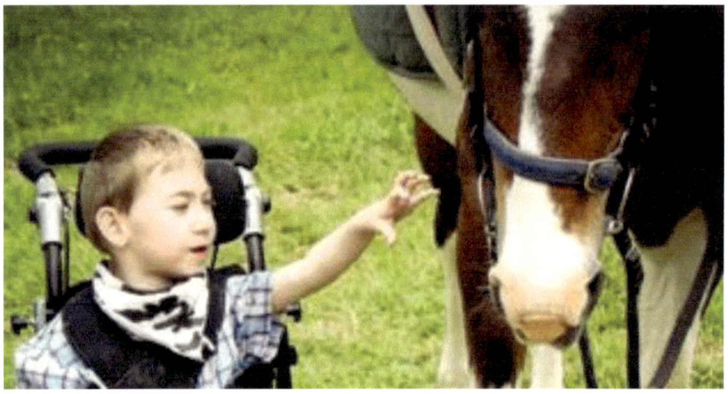

NZRDA offers advisory services, standardised training, and guides to the operational practices of its member groups. The RDA vision is: enhancing quality of life for those living with physical, intellectual, emotional and social changes. All RDA Groups are entities in their own right and offer a variety of services for their riders. RDA Groups offer individualised programmes according to the needs of their riders, but in general the following fields are covered by the RDA movement in New Zealand:

- Education- for riders with special needs, equine vocational training for high school students, as well as education and training for helpers, coaches and therapists.
- Therapeutic riding
- Rehabilitation
- Hippotherapy- NZRDA also offers qualifications for therapists and Horse handlers
- Interactive vaulting
- Psychotherapeutic riding
- Carriage driving, sport and recreation including support and assistance for Para-Equestrians, and Special Olympic riders.

Österreichische Initiative – Pferde helfen Menschen

An der Niederhaid 4
A-1140 Vienna
AUSTRIA
Contact: Gundula Hauser
Phone: 43 699 11 555 884
Email: g.h.hauser@gmx.at
Website: www.pferde-helfen-menschen.at

The *Austrian Initiative – Horses help Humans* offers advanced training for instructors and therapists, in the field of

- Equine assisted therapy (or THR) by "systemic pedagogic".
- Remedial-educational vaulting for students with special needs (mental retardation, awareness deficit).
- Counselling with the help of the horse.
- Competitive riding for riders with disabilities after the rules of Special Olympics.
- Equine assisted Learning - special body awareness through body language.

Tuition is charged for the courses and the courses are usually given in German, sometimes in English. A certificate is always issued at the conclusion of the course.

Österreichisches Kuratorium für Therapeutisches Reiten

Dambachstr. 74
Garsten 4451
AUSTRIA
Contact: Isabelle Boitllehner
Phone: 431 877 6725
Fax: 431 877 6725
Email: office@oktr.at
Website: www.oktr.at

The Austrian Committee for Therapeutic Riding is a non-profit organization chartered with the development and expansion of therapeutic riding, rehabilitative vaulting/riding and riding for the disabled in Austria. Within each field clients are holistically addressed: physically, emotionally, mentally and socially. The organization's goals focus on activities in the following areas:

1. Science
2. Basic and advanced training
3. Therapy
4. Rehabilitation
5. Horsemanship
6. Organization
7. The sport of riding and driving

The organization's goals are supported by the following means:

1. Development of guidelines for the training of, among others: physicians, physiotherapists, psychotherapists, psychologists, remedial and occupational therapists
2. Publication of a newsletter
3. Coordination of scientific symposiums
4. Coordination of advanced training seminars
5. Coordination of training courses

PACTH - Pennsylvania Council on Therapeutic Horsemanship

346 Briar Lane
Chambersburg, PA 17201
USA
Contact: Ann O'Shallie
Phone: 717 552-1628
Email: aoshallie@comcast.net
Website: www.pacth.org

PACTH's mission is to create awareness of the benefits horses can provide children and adults who have disabilities and to encourage the development of services for programs offering all types of equine assisted activities through education, research and application.

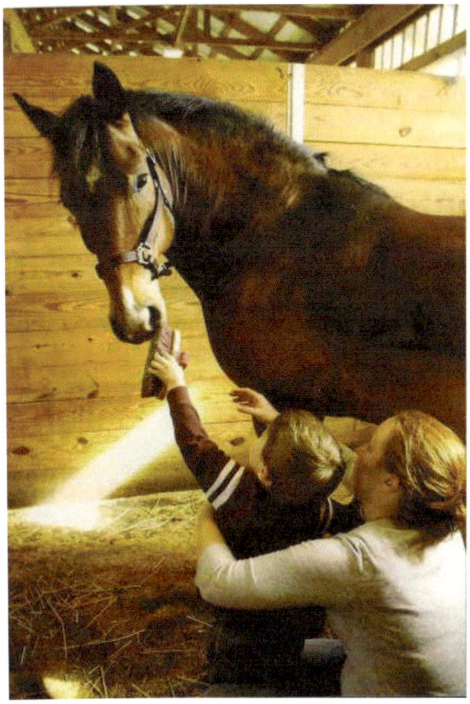

The Pennsylvania Council on Therapeutic Horsemanship (PACTH) is a coordinating council serving therapeutic horsemanship programs and the participants, instructors, volunteers and administrators who are involved in these programs.

Founded in 1984 by Penn State University, Pennsylvania State 4-H and the Pennsylvania Easter Seal Society, the PACTH has a long history of support to the profession of therapeutic horsemanship through the following:

- Workshops & Seminars
- Pennsylvania Qualified Instructor Training Course (PQI)
- Annual Conference
- Networking opportunities
- Start-up Information
- Newsletter
- Exhibits at Public Functions
- State Host for PATH INTL. Region II Conference

PATH International – Professional Association of Therapeutic Horsemanship International

PATH INTERNATIONAL
Professional Association of Therapeutic Horsemanship International

P.O. Box 33150,
Denver, CO 80233
USA
Contact: Kaye Marks
Phone: 303 452-1212
Email: kmarks@pathintl.org
Website: www.pathintl.org

The Professional Association of Therapeutic Horsemanship International (PATH Intl.) was founded in 1969 as the North American Riding for the Handicapped Association to change and enrich lives by promoting excellence in equine assisted activities. Since its formation the association is no longer limited to North America; members provide many services in addition to therapeutic riding; and an infinite number of special needs, thankfully no longer thought of as handicaps, are known to benefit from the power of the horse.

With over 800 member centers and certified professionals around the globe, 48,000 children and adults find physical and mental well-being plus a sense of independence through involvement with equines. In addition to therapeutic riding, a center may offer any number of equine assisted activities including hippotherapy, equine facilitated mental health and psychotherapy, driving, interactive vaulting, trail riding, competition, ground work, and stable management.

Through a variety of educational resources and member benefits (including an annual educational conference which attracts 800 attendees from around the globe) the association helps individuals start and maintain successful programs. With more than 43,200 volunteers, 3,800 Certified Instructors and professionals, 6,500 equines and thousands of contributors from all over the world PATH Intl. inspires and enriches the human spirit.

PHA - Polish Hippotherapeutic Association

100 Nowoursynowska St.
02-797 Warsaw
POLAND
Contact: Anna Struminska
Phone/Fax: 48 22 59 31 900
Email: biuro@pthip.org.pl
Website: www.pthip.org.pl

The Polish Hippotherapeutic Association (PHA) was founded in 1992 and has been a member of HETI since 1999. The Society was established by doctors, therapists and horsemanship instructors who since the 80's engaged in introducing hippotherapy into the rehabilitation of people with disabilities, especially for children with cerebral palsy. In Poland, the word hippotherapy defines all therapeutic activities associated with horses and riding, therefore PHA offers not only physiotherapy with the horse but also therapeutic riding and even recreation and horsemanship sports for people with disabilities.

The mission of PHA includes:

- supporting and promoting hippotherapy as a rehabilitation method
- supporting and promoting hippotherapy as beneficial within a pedagogic, psychotherapeutic and sociotherapeutic context
- improving and unifying hippotherapists' skills
- providing members access to professional literature and information about hippotherapy's progress in Poland and abroad

The Association promotes the best hippotherapeutic centers by supporting them with their patronage and with individual organization or by lending patronage to events of hippotherapeutic organizations, such as conferences and training sessions.
Specialists at PHA established a training program for hippotherapists and issued professional qualifications. 1612 people have been trained by that program since its inception in 1994. Experienced therapists from the Association also developed a set of basic rules of conduct for hippotherapists, the Canon of Polish Hippotherapy. Furthermore, PHA publishes its own magazine, Hippotherapeutic Review, to promote professional, safe and responsible hippotherapy.

RDAA - Riding for the Disabled Association of Australia Ltd

P.O. Box 280
Kent Town, SA 5071
AUSTRALIA
Contact: Ross Goulden
Phone: 61 (0)8 8364 4444
Fax: 61 (0)8 8364 4455
Email: admin@rda.org.au
Website: www.rda.org.au

Riding for the Disabled Association Australia (RDAA) is a nonprofit organisation which provides opportunities for anyone with a disability to enjoy safe, healthy, stimulating, therapeutic, horse-related activities. RDAA develops ability and enriches lives through specialised programs and relationships with horses. RDAA:

- is the federal body of 8 autonomous RDA member states and territories in Australia;
- acts as a national body and resource for its member states and territories to access accreditation, education and insurance;
- encourages national links between the 8 affiliated states and territories offering therapeutic riding, driving and equine related activities;
- is actively involved in establishing national policies and standards to ensure excellence in service delivery of RDA programs across the country;
- liaises with government and industry on national issues;
- leads in equine matters related to therapeutic riding;
- plays a significant role as a member of the Federation of Horses in Education and Therapy International (HETI)
- promotes elite athletes at all levels
- 5 RDAA riders represented Australia at 2008 Paralympics

From its mission statement: "RDA enables people with disabilities and volunteers throughout Australia to experience enjoyment, challenge and a sense of achievement through participation in equestrian activities, resulting in the development of life skills, improved quality of life and attainment of personal goals. We are dedicated to providing these services through an organisation that is relevant, innovative, effective and rewarding. Operated by trained and caring people committed to our aims now and in the future, RDA is recognised as having active involvement within the equestrian and general communities."

HONG KONG

馳騁

RDA - Riding for the Disabled Association Hongkong

Lot 45
Lung Mun Rd.
Tuen Mun, N.T.
HONG KONG SAR
Contact: Yumi Toda
Phone: (852) 2454 9961
Fax: (852) 2465 5914,
Email: info@rda.org.hk
Website: www.rda.org.hk

Riding for the Disabled Association Ltd. (RDA) was founded in Hong Kong in 1975 by the then Governor's wife, Lady MacLehose. A non-profit organization, RDA provides riding instruction and other associated activities to adults and children with disabilities. Their aim is to teach these members of their community to ride a horse to the best of their ability and to learn to interact with the horse and each other. Currently, RDA has 20 horses and ponies. Each year, it offers 28 free weekly sessions for 36 weeks at 3 RDA Centers in Pokfulam, Tuen Mun and Lo Wu. Each session lasts for 45 minutes to an hour and accommodates 6 riders. In all, RDA offers in excess of 6,000 lessons per year. In addition there are lessons run for larger groups so that schools and other organizations can decide if riding would be beneficial.

Competitive riding is a more recent addition to RDA's offerings. Late in 2005, the Sports Sub Committee of the RDA General Committee was formed with a mission to find and train Para Equestrians for competition in local, regional, international and the Paralympics. Currently, RDA has identified a few young athletes who have potential and provides regular training to these young athletes, working towards having them compete in local shows and hopefully international competitions in the future.

RDA - Riding for the Disabled Association Ireland

167 Barrington Court,
Prosperous
Co. Kildare
IRELAND
Contact: Niamh Kingston
Phone: 353 1 387 6498
Email: rdaisecretary@eircom.net
Website: www.rdai.org

RDA Ireland, founded in 1968, is a voluntary organisation providing opportunities for riding or carriage driving to people with intellectual or physical disabilities. At present 70 groups (3 carriage driving) give weekly riding or driving lessons for approx 1000 children and adults around the country. Most groups work on weekdays in riding schools, whose owners very generously allow the use of their arenas and well schooled horses and ponies. Each group has its own committee, an approved instructor and as many helpers as required. The Council co-ordinates all this activity, arranges insurance, fundraising, and publicity.

The Training Committee runs local and National Training Days for helpers, instructors and therapists – always very well attended. Instructor Assessments are also held annually for group instructors who have completed a minimum of 30 hours teaching. Carriage drivers compete locally and at such prestigious National events as Punchestown and Dublin International Horse Show. They also travel to compete internationally including European and World Championships.
In addition to weekly riding lessons riders have the opportunity to compete in Special Olympics – Ireland hosted the World Special Olympics in 2004- and to participate in special classes and demonstrations in local shows. The highlight of the year is the RDS - venue for the Dublin International Horse Show in August. Application forms are sent out to all groups in early Spring and the final selection sees 8 lucky riders and their mounts from around the country arriving at Fairyhouse for a Trail Ride practice at an Irish Pony Society Show. On the big day in Dublin, each rider has a large following of parents and friends and group members as they ride into a large arena with their leaders and helpers. There is always a big audience, a photographer and a celebrity to present rosettes. Many of the riders become celebrities in their own communities, with coverage and photos in local newspapers.

RDA - Riding for the Disabled Association Japan

36-7-103 Sakashita 1-Chome
Itabashiku, Tokyo 174-0043
JAPAN
Contact: Ms. Miho Kaneko
Phone: +81 80 6519 7818
Email: rda@rdajapan.com
Website: www.rdajapan.com

RDA Japan is a certified non-profit organization which promotes and supports safe riding and activities with the horse for people with disabilities in order to improve their quality of life. For this purpose, they

- organize workshops and seminars on therapeutic riding and equine issues, as well as on welfare of horses
- offer training courses for instructors, volunteer helpers and therapists
- support unit-groups to improve their activities
- assist new groups in their start-up phase
- send qualified instructors to RDA-related activities
- provide support to groups through our network of therapists
- engage in international exchange
- promote awareness of the general public on equine-assisted activities

At the 2010 unit meeting, the members will come together to share their experience and discuss topics of common interests. This year's main theme is safety.

For the second half of the year the Instructor's Certificate Test is being planned and RDA looks forward to welcoming new instructors.
Another important task this year is to strengthen the ties with other organizations and cooperate in promoting therapeutic riding in our country.

RDA-KRA - South Korea Racing Authority

685 Juam-dong
Gwacheon-si
Gyeonggi-do 427-711
SOUTH KOREA
Contact: Ms Shin Jung Soon
Phone: 822 509 1695 82250
Email: 88hada@hanmail.net
Website: http://company.kra.co.kr/global/english/main

KRA joined HETI as a full member in August 2010 after three years of being an Associate member. The Korea Racing Authority (KRA) is a non-profit organization established by the KRA Law. It is the sole racing authority in Korea and is under the supervision of the Ministry for Food, Agriculture and Fisheries.

KRA's objectives are to authorize, operate horse racing and promote the equestrian industry in KOREA. One of the main objectives is to be a leading influence in charities and community services. Riding for the disabled is the main community service of KRA.

KRA started RD programming in 2005. At that time it was a pilot project because KRA had no access to a professional RD instructor or advisor here in Korea. Now, KRA has 2 RD instructors and one of them is also an Occupational Therapist as well. In 2010 they are providing hippotherapy and therapeutic riding programs with 5 therapeutic riding ponies for 108 clients.

In addition, the KRA strives to be a leader in disseminating RD in Korea. KRA has hosted several PATH INTL. workshops & certification courses in the last 3 years. Many people are interested in this field and KRA wants to ensure that RD instructors have access to knowledge, practical experience and safety awareness.

RDA - Riding for the Disabled Association UK

RDA It's what you can do that counts

Riding for the Disabled Association
Incorporating Carriage Driving

Norfolk House
1 Tournament Court
Edgehill Drive
Warwick CV34 6LG
UNITED KINGDOM
Contact: Mrs. Julie Dargie
Phone: +44 (0)845 658 1082
Email: info@rda.org.uk
Website: www.rda.org.uk

RDA has spent 40 years delivering opportunities for therapy, achievement and enjoyment to people with disabilities. Founded as a national organisation in 1969, RDA now has 500 member groups across the UK – together, these groups help 28,000 people to take part in riding, carriage driving and vaulting activities.

Although set up to work with people with physical disabilities, RDA Groups now work across the spectrum of disabilities, as well as age range, social status and urban and rural environments. The focus of RDA's work is to ensure that each individual derives a direct therapeutic benefit and has the opportunity to achieve their personal goal (whether that be learning a very simple skill or winning a major competition).

RDA gives clients the opportunity to:

- Reach therapeutic goals
- Achieve their personal ambitions
- Combat social isolation
- Develop life skills
- Experience the outdoors
- Connect with animals

Although the current level of service delivery is impressive, there are many more people who merit help and support. Currently, RDA does not fully meet the increasing demand as the vast majority of RDA Groups are reporting a waiting list. RDA's focus, therefore, is on increasing capacity to serve even more clients.

SARDA - South African Riding for the Disabled

P.O. Box 235
Constantia, 7848, Western Cape
South Africa
Contact: Belinda Thom-Bain
Phone: 27 82 449 5176
Email: national@sarda.co.za
Website: www.sarda.co.za

South Africa is a beautiful and diverse country; 1,2 million square kilometres in size with a population of about fifty million. Well known for its sunny climate, magnificent scenery and of course Madiba (Nelson Mandela), this year South Africa was the first ever African nation to host the soccer world cup.

South Africa's population is multicultural and multiracial and the 400 plus riders that visit SARDA's six facilities each week are drawn from all groups. To ensure that riding is available to all there is no charge for any of the sport riding and riding skills classes offered. Despite a rigorous instructor qualification process SARDA's twentyfive instructors, along with many helpers, work on a voluntary basis. SARDA employs only eight staff members.

Development of PARA riders falls under SARDA's umbrella as well and in November 2009, the Cape Town branch hosted the first ever FEI 3* event held in South Africa. Thanks to a generously awarded scholarship, SARDA is now also instrumental in the setting up and running of an Equine Assisted Activities & Hippotherapy Association.

As a Third World country South Africa has a small tax base so no government financial support is available. SARDA is a registered non-profit organisation and keep their riders in the saddle thanks to fund raising efforts, generosity of the public and sympathetic donors.

SRT - Association of Finnish Riding Therapists

Välitalontie 11
87850 Paltaniemi
FINLAND
Contact: Sanna Mattila-Rautiainen
Phone: 358 41 454 2428
Email: sanna.mattila@saunalahti.fi
Website: www.suomenratsastusterapeutit.net

The Association of Finnish Riding Therapists - SRT was founded in 2000. It serves as a link between its members, attends to their interests and the interests pertaining to the field of riding therapy in general. One of SRT's goals is to help its members to continually improve their professional skills and knowledge through courses and lectures. The association provides quidelines to the quality, ethics and practice of hippotherapy in Finland. They have created an active national and international cooperation with other organizations that share related interests.

A training program started in Finland in 1988. The training is completed in approximately three years. The entrance requirements to the training are a college degree in a health care profession, pedagogy or social sciences and two years work experience in the field of rehabilitation after graduation. Good horsemanship and riding skills are required. As part of the training program in 2013 a new course starts in Hevosopisto Equine College, carried out with University of Turku and supervized by the Association of Finnish Riding Therapists.

SRT has also created an active national and international cooperation with other organizations that share related interests in therapeutic riding. The Association of Finnish Riding Therapists has been a full member of HETI since 2002. 2010-2012 it was a silent partner to EU funded Hippotherapy- Shared Learning Opportunities created between Hevosopisto Equine College with ACPTR U.K. and the Polish Hippotherapeutic Association.

SV-HPR - Schweizerische Vereinigung für Heilpädagogisches Reiten

Bergstrasse 77
5430 Wettingen
SWITZERLAND
Contact: Luz Sozzi
Phone: +41 (56) 427 14 64
+41 (79) 442 95 58
Email: praesidium@sv-hpr.ch
Website: www.sozzi-brunner.ch

The Swiss Union for Therapeutic Riding (Schweizerische Vereinigung für Heilpädagogisches Reiten - SV-HPR) was founded in 1985 to promote among experts and the public the recognition of therapeutic riding. Main tasks of the SV-HPR are:

- Coordinating and supporting the activities of its members and representing their interests.
- Offering the training of riding therapists, certified by the SV-HPR (approx. 1000 Hours)
- regulating the training of riding therapists with the federal professional recognition as the goal.
- Promote training courses for its members in the field of horse-assisted activities and handling of horses.
- Promoting animal training and breeding of horses for therapeutic use. To that end the SV-HPR awards a plaque to quality model stables in recognition of their work in the service of people with disabilities.

SV-HPR is a member of the Federation of Horses in Education and Therapy International (HETI).

TRAG - Therapeutic Riding Association of Greece

P. Kanellopoulou 10
Goudi
11527 Athens
GREECE
Contact: Dimitra Karouzaki
Phone/Fax: 0030 210 7488566
Email: triding@otenet.gr
Website: www.trag.gr

The Therapeutic Riding Association of Greece (T.R.A.G.) was founded in 1992 as a voluntary and non-governmental organization. A year later, it acquired its athletic identity. In 2004, TRAG obtained its own center of riding for people with specials needs in Goudi, Athens.

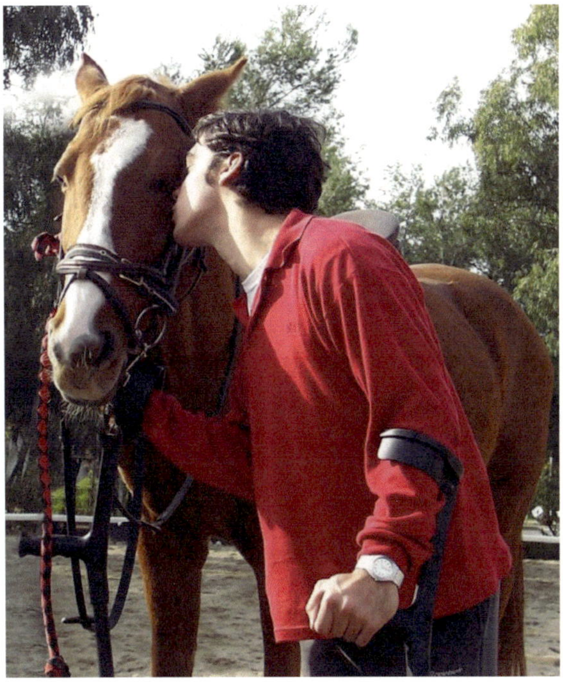

The workforce of the Association consists of six people and eight horses and the Association offers services to over 100 clients weekly. This wouldn't be possible without the help of more than 50 volunteers (Greek and EVS volunteers from different countries) and of course many known and unknown sponsors who have been supporting the Association from its very beginning.

Within the bounds of increasing progress and improvement, T.R.A.G. has been trying to help form professionals who have knowledge and sensitivity for people with disabilities. For this reason, seminars on the topic of Therapeutic Riding are constantly organized, in order to disseminate the idea of Therapeutic Riding and further its acceptance as a means of therapy.

In 2012, the XIVth International Congress of Therapeutic Riding will be held in Athens, Greece. T.R.A.G is the proud host and has already started the preparations for this important event that will contribute to the promotion and expansion of Therapeutic Riding all over Greece. For more information regarding the Congress visit www.frdi-athens2012.gr.

Turkish Equestrian Federation

Poligon Caddesi No 4
Sarıyer / Istanbul 3446
TURKEY
Contact: Ms. Yasemin Inan
Tel: 90 532 277 2057
Email: yaseminan@hotmail.com
Website: www.binicilik.org

The Turkish Equestrian Federation has a PATH Intl. certified instructor giving adaptive riding lessons in one of their clubs. Under the project name 'My horse,my wing', they are working hard to apply the regulations and rules of adaptive riding and hippotherapy based on international standarts to raise new riders for the paraolympics.

They have signed a protocol for the project with one of their sports clubs on the 3rd of December, the International Day of People with Disabilities. The name of this sports club is Galatasaray and the website of this club is www.galatasaray.org. Galatasaray is particularly important for the federation because the club is their first sponsor for the project. The club is particularly well known for its football team, but as a large club it is dealing with many areas of sports including riding. To that end, the president of the club had his picture taken on that day with the children and horses to demonstrate the importance of adaptive riding and hippotheraphy.

Warashibe-kai Social Welfare Organization / Warashibe Therapeutic Riding Center

120-2 Nishicha, Urakawa-cho
Urakawa-gun, Hokkaido
057-0171
JAPAN
Contact: Yuki Chiba
Tel: 0146 24 8221
Fax: 0146 24 8222
Email: ryoiku@cocoa.ocn.ne.jp
Website: www.jyohba-ryoiku.or.jp Facebook: ja-jp.facebook.com/warashibehorse?filter=1

The Warashibe Social Welfare Organziation has 3 different places and type of facilities. They established the Therapeutic Riding Center which is attached to the Urakawa Warashibe Residential Facility for the Disabled. Therapeutic riding at Warashibe-kai is called “riding conductive education”. It provides people with mental and physical disabilities an opportunity to touch and be with horses and to ride horses, as well as a chance to maintain and improve their abilities while enjoying themselves. Therapeutic riding is also used to promote participation in society through horseback riding.

We are using various kinds horses from 4 to 24 years old with a lot of experience. The riders are enjoying therapeutic riding from 2 to 92 years old. For the elderly people, therapeutic riding is used for the purpose of preventing the need for long-term care. Therapeutic riding instructors and physical therapists are involved with the riding lesson.
We are also researching the effects of Therapeutic Riding:

- Changes in pressure distribution on seat before and after riding therapy
- Considering the mental effects of riding therapy
- Effects of riding therapy - example of application to the physical therapy of a subject with partial paralysis on the right side.

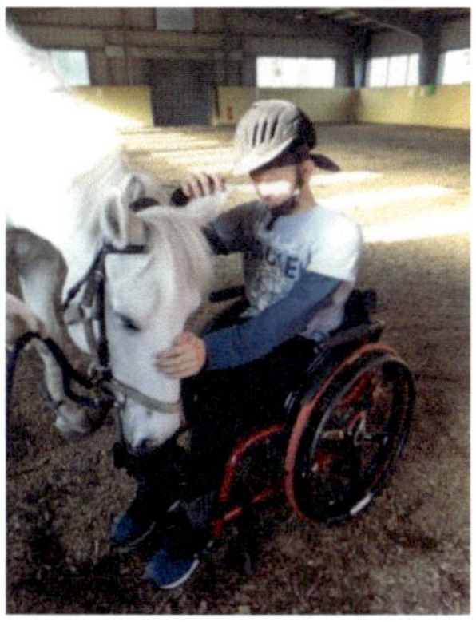

2012 HETI Annual Membership Directory

Associate, Concession and Honorary Members

Associate Members

A.V.R.E.S., ITALY
Able the Disabled Foundation, THAILAND
Abrams, Barbara, USA
Adams-Wheeler, Sue, UK
AIASPORT ONLUS, ITALY
Ann Sands O.A.M., AUSTRALIA
APRES, ITALY
Arteveldehogeschool, BELGIUM
Ass. Reg. D'Equit. Therap. D'Alsace, FRANCE
AWO Dortmund, Henrike Struck, GERMANY
Barbara Engel Therapy Services, USA
Barnfield, Anne, CANADA
Barry, Leora, USA
Beitostolen Helsesportsenter, NORWAY
Benjamin, Joann, USA
Boyce (Rufus), Sharon, South Africa
Brophy, Mary, IRELAND
Brown, Octavia, USA
Bruna Grubits Freire, Heloisa, BRAZIL
Butler, Sharon, USA
Buyse, Brigitte, BELGIUM
Cabre, Geni, UNITED ARAB EMIRATES
Carlisle Academy, USA
Cavalo Amigo, PORTUGAL
Centre Equestre Equovita, BRAZIL
Centro de Rehab. Infantil HOGA, MEXICO
Claremont TR Centre, AUSTRALIA
Corte-Real, Racquel, PORTUGAL
Crnic, Tamara, SLOVENIA
CZECH TR Association, CZECH REPUBLIC
Dalia Kreindler Assoc., ISRAEL
De Acuna, Fernanda Morey, PERU
Denby, Jennifer and Douglas, AUSTRALIA
Donahue, Jen, USA
Duff, Beth, UK
Equest, USA
Equinesport Emotional Management, SPAIN
Equipa d Equitacao Psico, PORTUGAL
F.Q.E.T, CANADA
Faubel, Giselle, USA

Associate Members (continued)

Festina Lente Foundation, IRELAND
Field of Dreams Foundation, UK
Fitzpatrick, Lurene, IRELAND
Foerderkreis fuer Therap Reiten, GERMANY
Francis, Sally , AUSTRALIA
Fulcrum Therapy, USA
Fundacio Terapia a Cavall, SPAIN
Garrigue, Rene, FRANCE
Gliedman-Baker, Shari, USA
Goeteborgs Handikappridklubb, SWEDEN
Goetske, Birgitte, DENMARK
Grannell, Marian, UK
Green Chimneys, USA
Gross-Naschert, Edith, MEXICO
GSATRH, GREECE
Haartz, Elizabeth, USA
Hakanson,Margareta, SWEDEN
High Hopes Therapeutic Riding, USA
High Hopes, USA
Hipoterapi Antalya, TURKEY
Hippotherapy RDA WA, AUSTRALIA
Hoff, Jean Paul, FRANCE
Hoffmann,Gerlinde GERMANY
Houry, Judy, SAUDI ARABIA
IAPE EQUO, BRAZIL
Ihara, Masayo, JAPAN
Iversen, Hanna Dahl, DENMARK
Kamizi, Kallirroi, GREECE
Kay, Judi, ISRAEL
Klouman, Marianne, NORWAY
KRILA- terapijsko jahanje, CROATIA
Kulichova, Jana, CZECH REPUBLIC
Kulmala, Tanja, FINLAND
Lanark County TR Program, CANADA
Laurie, Harriet, UK
Liga, Maria, GREECE
Lightsey, Priscilla, CHINA
Lipscomb, Pauline, UK
Liu, Ming, CHINA
Longden, Mary, AUSTRALIA

Associate Members (continued)

Lysell, Jeanette, NORWAY
Macauley, Beth, USA
Malisia, Andreia, CANADA
Mane Stream, USA
Masatoshi, Nakata, JAPAN
Mastorou, Perdika, GREECE
Mattila-Rautiainen, Sanna, FINLAND
McIntyre Centre, AUSTRALIA
Mehlem, Monika, GERMANY
Mona I. Thelle, NORWAY
Morris,Teresa, USA
Municipeo de San Nicolas, MEXICO
Nakai, Yukari, JAPAN
Nakata, Masako, JAPAN
Northern Lights TR Association, CANADA
Orde, Samantha, UK
O'Shallie, Ann, USA
Pal-O-Mine Equestrian, USA
Paradis, Lee, USA
Pauline Perry O.A.M., AUSTRALIA
Pearce, Norma AUSTRALIA
Pegasus RDA of the ACT, AUSTRALIA
Petriki, Petroula, GREECE
Plaza, Alison, USA
Promitheas A.M.K.E, GREECE
RD Samsung, South Korea
RDA Australia Hippotherapy, AUSTRALIA
RDA (NSW), AUSTRALIA
RDA of Dubai, UNITED ARAB EMIRATES
RDA SA Inc, AUSTRALIA
RDA SINGAPORE, SINGAPORE
Renker, Lorrie, USA
Rindfleisch-Wu, Uta, TAIWAN
Sayler, Pat, USA
Schoerle, Armgard, GERMANY
Schulz, Marietta, GERMANY
Shkedi, Anita, ISRAEL
Silkwood-Sherer, Debbie, USA
Sistema Municipal DIF de Puebla, MEXICO
SOD ITALIA, ITALY

Associate Members (continued)

Sokolof, Marilyn, USA
Sosiaaliped. Hevostoimintayhdistys, FINLAND
SPANA, MALI
Spirit Open Equestrian Program, USA
Splinter-Watkins, Kathryn, USA
St. Andrews Presbyterian College, USA
Strides to Success, USA
Sunshine Coast Riding, AUSTRALIA
Sweeney, Molly, USA
Touchstone Farm, USA
TR Center in Ioannina, GREECE
Tuck, Susanne, AUSTRALIA
UNH Therapeutic Riding, USA
Vetsch, Danielle, FRANCE
Walker, Carolyn, UK
Webb, Olive, NEW ZEALAND
Wheelwright, Christine, USA
Wilson College, USA

Concession Members

Grey Horse Learning, AUSTRALIA
Redmond, Janalee, USA
Tanaka, Ayako, JAPAN
Cormier, Genevieve, CANADA
Davey, Daphne, CANADA
Poole, Joanne, CANADA
James, Jane, CANADA
Mac Kinnon, Joyce, CANADA
Sponsel, Lelia, CANADA
Statelli, Alessandra, ITALY
Skyrud, Beate, NORWAY
Kern-Godal, Ann, NORWAY
Brown, Alexandro, SCOTLAND
Baillie, Jenny, UK
Nelson, Yvonne, UK
Siddens, Nancy, USA
Tebay, Jean, USA
Mantell, Laura, USA
Ulaszek Benjamin, Mina Nicolle, USA
Anderson, David, USA
Love Quest Children's Foundation, USA

Honorary Members

Ernst Reinhardt Verlag, GERMANY
Horses and Human Research Foundation, USA
IAHAIO, INTERNATIONAL ORGANIZATION

2012 HETI Annual Membership Directory

Index of Members by Country

Index of Members by Country

Australia

Austria

Belgium

Belgium (cont'd)

Brazil

Canada

Index of Members by Country (continued)

Index of Members by Country (continued)

Index of Members by Country (continued)

Luxembourg

ATE - Association Therapie Equestre (Full Member)

Mali

SPANA (Associate Member)

Mexico

AMET - Mexican Association of Therapeutic Riding (Full Member)
Centro de Rehabilitacion HOGA (Associate Member)
Edith Gross-Naschert (Associate Member)
Municipeo de San Nicolas (Associate Member)
Sistema Municipal DIF de Puebla (Associate Member)

Netherlands

FPG - Federatie Paardrijden Gehandicapten (Full Member)

New Zealand

NZRDA - New Zealand Riding for the Disabled Association Incorporated (Full Member)
Olive Webb (Associate Member)

Norway

Ann Kern-Godal (Concession Member)
Beate Skyrud (Concession Member)

Norway (cont'd)

Beitostolen Helsesportsenter (Associate Member)
FTR - Faggruppen for Terapiridning (Full Member)
Hest og Helse (Full Member)
Jeanette Lysell (Associate Member)
Marianne Klouman (Associate Member)
Mona Thelle (Associate Member)

People's Republic of China

Ming Liu (Associate Member)
Priscilla Lightsey (Associate Member)

Peru

Fernanda Morey Yrigoyen de Acuna (Associate Member)

Poland

PHA - Polish Hippotherapeutic Association (Full Member)

Portugal

Cavalo Amigo, Portugal (Associate Member)
Equipa d Equitacao Psico (Associate Member)
Raquel Corte-Real (Associate Member)

Republic of China (Taiwan)

CTEA - Chinese Taipei Equestrian Association (Full Member)
Uta Rindfleisch-Wu (Associate Member)

Index of Members by Country (continued)

Index of Members by Country (continued)

Index of Members by Country (continued)

USA (cont'd)

Worldwide